This book is dedicated to all who find Nature not an adversary to conquer and destroy, but a storehouse of infinite knowledge and experience linking man to all things past and present. They know conserving the natural environment is essential to our future well-being.

RAINBOW BRIDGE
THE STORY BEHIND THE SCENERY®

Text and Photography by Gary Ladd

Gary Ladd has been a photographer and author for 25 years, thoroughly exploring the canyon country of the Southwest. He first photographed Rainbow Bridge in 1975 and rowed his dory the length of Lake Powell in 1976. Having backpacked to Rainbow Bridge dozens of times, he thinks of the monument as his backyard.

Rainbow Bridge National Monument, *located in southern Utah, was established in 1910 to preserve the world's greatest known natural bridge and a portion of the wilderness around it.*

Front/back covers: Rainbow Bridge, Earth's largest natural bridge. Inside front cover: Courtesy docks lead boaters to Rainbow Bridge viewing areas. Page 1: Globe mallow blooms. Pages 2/3: Visitors are dwarfed by Rainbow Bridge.

Edited by Cheri C. Madison. Book design by K. C. DenDooven.

*B*alance, improbability, grace,
proportion, simplicity,
and the aura of beauty that flows from mystery,
combine in Rainbow Bridge to form
a masterwork of Mother Nature.

*N*estled in the surrealistic sandstone landscape of southern Utah is a tiny national monument embracing only 160 acres. It is a dot lost in a chaos of slickrock canyons, cliffs, slots, pinnacles, folds, towers, terraces, cracks, and domes. But within that speck resides one of the giants of the National Park System—Rainbow Bridge! With a span of 275 feet, a height of 290 feet, and a soaring figure that breaks the heart with its purity, it is the largest natural rock span on Earth.

To explore Rainbow Bridge's natural and human history, we must embark on two journeys, the first into geologic time, the second into early 20th-century time.

Be advised that Rainbow Bridge is a recent development in terms of geologic time. Therefore, much of our story occurs before there's a trace of the stone rainbow. Patience, immense patience in geological matters is essential. You might say that geologic time is not "abridged"!

So, here we go, back before there was a Lake Powell, back before the Colorado rumbled through its canyons, back before there was a hint of Rainbow Bridge, back before there was a Navajo Mountain, back, back, back....

Bridge Canyon corkscrews through the sandstone labyrinth at the foot of cloud-enshrouded Navajo Mountain. For decades Anglo civilization flowed around this wilderness fortress, effectively concealing Rainbow Bridge from the outside world.

*It is clear that Rainbow Bridge
is a work in progress,
and we are fortunate to stand here today
when it seemingly floats
in a state of consummation.*

Inklings of the Future

Whoa! Far enough! What time is it? Oh yes, it's a tick over 200 million years before the conventional present. We haven't gone far, only about 4 percent of Earth's history. But here, where we stand, is the future site of Rainbow Bridge. To cook up something like Rainbow Bridge, the first ingredient is the material from which the rock of the bridge is made.

QUESTIONABLE LANDSCAPES

The region that we know today as southern Utah and northern Arizona is absolutely unrec-

▲ **A crayfish prowls over 200-million-year-old** sandstone. The crustacean doesn't know it, but about 50 million years before this rock was deposited, life on earth was nearly exterminated by a mysterious catastrophic collapse of the ecosystem.

ognizable—it's a near-featureless flood plain where streams endlessly tumble sand from here to there and on toward the sea. The Earth is already very old—4.4 billion years.

Looking backwards in time from our stance in the stream, it has only been 250 million years since plants first ventured out of the teeming oceans to colonize the continents. Animals, attracted by the bounty of unharvested plants on the dry land, soon concocted methods themselves to survive in the alien atmosphere beyond the water.

Still looking backward in time it was only about 50 million years ago (250 million years before Lake Powell Time) that something terrible happened. Up to 90 percent of all Earth's species perished in an environmental catastrophe. Scientists call it the Great Dying but as to the cause of the disaster, they're baffled. Nevertheless, reptiles flourished in the aftermath of the Great Dying. Among them were the dinosaurs. They're here, somewhere nearby. Look: Footprints. Careful, you're stepping in one of them now.

KAYENTA SANDSTONE

Let's consider our stream bed in the light of a global view: Off to the north and east there are as yet no true Rocky Mountains, only some ancestral Rockies. Much farther east the Appalachian Mountains are youthful and rugged, far more impressive than the worn rolling mountains of future Lake Powell Time. The Appalachians were raised when the Eastern and Western hemispheres collided in a kind of continental fender bender. Beyond the mighty Appalachians lies Europe. There is no Atlantic

Maroon ledges of Kayenta Sandstone cradle Bridge Creek. *The geologist Herbert E. Gregory officially named the Navajo Sandstone in 1915. Ironically, Gregory was invited to join the 1909 search for Rainbow Bridge—perhaps the finest expression of Navajo Sandstone—but was unable to go along.*

Ocean. And off to the west, there is no California. Its parts have yet to dock with the continental margin.

Consider this: The continents (and the tectonic plates of which they are the highest parts) slowly slide around the surface of the earth. They have, presumably, always done this. They drift north and south, east and west, rotate, crash into each other, scrape against one another, coalesce and break apart. Currently (still stuck in our imaginary mud and sand), North America and Europe remain welded together. A dinosaur could saunter from New York to Paris. But it won't be this way forever. North America is uncoupling from Europe even as we speak.

With the plates wandering about between equator and pole, with mountains thrusting and eroding, with basins forming to receive the detritus from the mountains, any location will experience a variety of different environments given enough time. These stream-deposited sands will one day become known as the Kayenta Sandstone, solid evidence of an active stream and river system early in Jurassic time. Remember, there are no towering cliffs such as those that surround Rainbow Bridge in Lake Powell Time—just flat stream deposits. Memorize the Kayenta Sandstone's name and its muted purple complexion.

__R__ock layers are ▷ *remnants of past environments. The sands that eventually lithified into today's Navajo Sandstone accumulated in a vast windswept desert about 200 million years ago. The desert stretched from central Wyoming to southern Nevada, rivaling the size of today's Sahara Desert. In the vicinity of Rainbow Bridge the Navajo Sandstone is over 1,000 feet thick, nearly all of it composed of fine quartz grains. (In Zion National Park, the Navajo Sandstone reaches a thickness of 2,000 feet.) A pageant of environments succeeded the sandy desert, most of them leaving behind traces of their reign in the form of additional rock layers stacked upon the sandstone.*

*◀ **B**ridge Creek, meandering toward the Colorado River, often created notch-like shortcuts as the meanders widened and connected. These old curving meander paths are still evident today. But in one location, the creek punched a hole through its meander partition. The massive cliff above the break stood firm—and Rainbow Bridge was born.*

NAVAJO SANDSTONE

Change is in the air. In a few million years the streams go dry. The region passes into desert. Sand dunes march across the land. The dune fields cover much of the future western United States. Dinosaurs still prowl around and we'll trace their tracks in a rock layer destined to become known as the Navajo Sandstone. Take note of this stratum, too. Unlike the Kayenta, the Navajo layer is unusually thick, well over 1,000 feet here and over 2,000 feet in some other locations. It is salmon tan in color.

In 10 or 20 million years the desert will be replaced by yet another environment—shallow marine deposits and tidal flats. This, in turn, will be superseded by still another and another. Thus the strata mount as sand and silt and lime accumulate. Ages pass. The Kayenta and Navajo sands are compressed and cemented beneath many hundreds of feet of more recent deposits. They harden into bona fide rocks. Oceans come and go. Lagoons, streams, swamps, and deserts take their turns. Eons slip by.

Sporadic periods of uplift and erosion strip away portions of the uppermost strata. But it is not until about 5.5 million years before Lake Powell that uplift comes in a powerful spasm. After half a billion years of relative calm, the Colorado Plateau is rapidly elevated and dissected. A major canyon-walled river system develops. We know it as the Colorado River.

Now we walk on a landscape that looks vaguely familiar. A search of the area may suggest the exact site of the future Rainbow Bridge.

A little stream flows down from a dome-shaped mountain to the southeast. The stream follows a series of meanders cut into a salmon sandstone. The date is not well known. We'll call it several thousand to a few tens of thousands of years before Lake Powell Time. Just yesterday.

As we walk along the entrenched stream meanders, it's not obvious how close some of the loops come to one another. But let's climb up above the stream bed onto a bench from which we'll have a commanding view. Here, put on this hard hat. A few rocks may tumble from above before our journey is over!

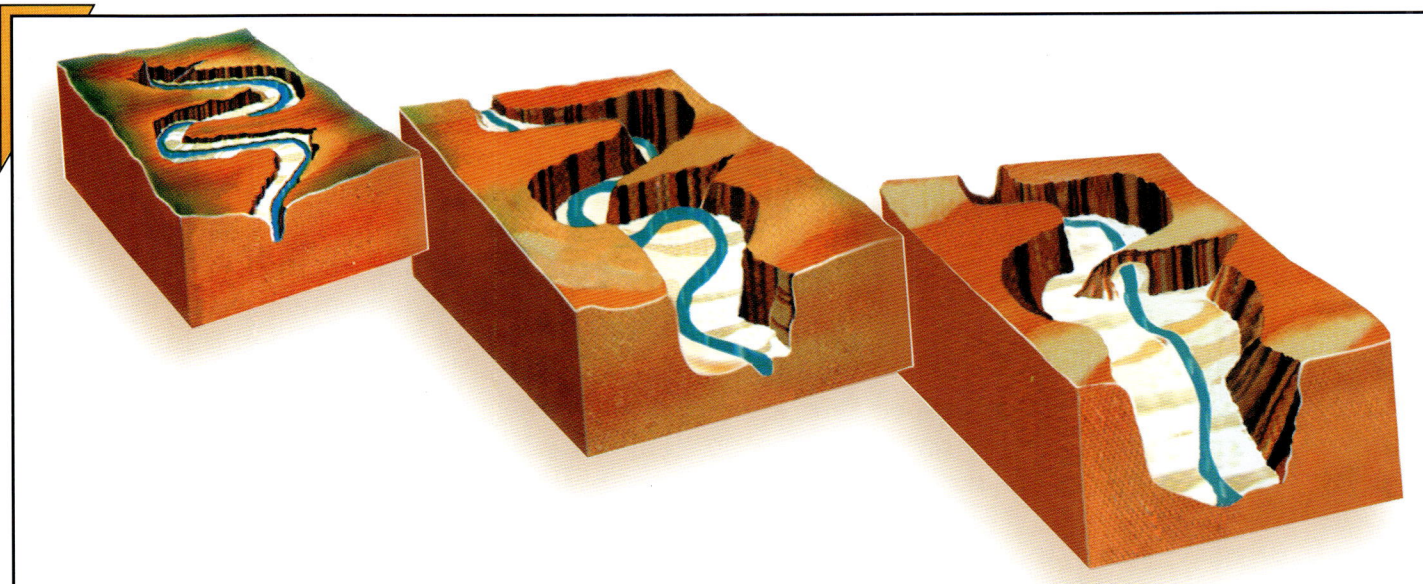

Bridge Creek, like all streams, exhibits a tendency to wander and develop meanders. As the stream carved its canyon, the meander loops broadened and eventually broke through many of the rock partitions that separated them. Sections of streambed were abandoned as these shortcuts were created. But at one site, a rock window formed where Bridge Creek broke through the partition without toppling the breached cliff. Over many millennia, floods, gravity, and weathering enlarged the window and shaped the massive rock above it into Rainbow Bridge.

◁ ***Here are the tools which first created***
Rainbow Bridge: rocks, cobbles, pebbles, and grit, swept downstream by floods. They pummeled and pounded away at the stream's curving alcove turns. Eventually, working from both sides, they battered their way through the thin partition of rock to daylight.

At that moment a section of the thinning rock partition between the loops breaks down, and the raging stream roars through the shortcut. The shortcut quickly widens and prospers while the loop is abandoned.

The rock here is Navajo Sandstone. The former overlying rock has already been stripped away. But as we watch, a new layer is beginning to appear, a purplish stratum. Sound familiar? It's the Kayenta Sandstone.

Because the Kayenta is a little more resistant than the overlying Navajo, downcutting is retarded somewhat. But the floods still come; the loops still broaden. But now they intersect more frequently near the Kayenta-Navajo contact.

The rock partitions breach and collapse. The stream sweeps away the rubble. But watch what happens at that last loop. The floods rasp at both sides of the fin, cutting deep alcoves as we've seen elsewhere. Finally a breakthrough occurs. But this time the fin, more massive than most, does not collapse. Instead, the stream rushes not through a notch but beneath a stone bridge!

At first the opening is only a few feet across. And yet this is the birth of Rainbow Bridge.

Keep watching. Bridge Creek persists in its work and enlarges the hole by downcutting and widening. Meanwhile, pieces of rock fall from the underside of the bridge. The form is perfected by the action of frost-wedging, gravity, weathering, and a generous dash of time.

We must also remember that the local climate changes over time and that the creation and development of Rainbow Bridge may have been influenced by meteorological conditions different from those with which we are familiar.

Charles L. Bernheimer, who first visited the bridge in 1920, grasped the process and wrote: "...the evolution of the ages have merely brought to the surface its muscular structure, divesting it of weak and useless particles."

And now we find ourselves in the present, safe and sound. It is clear that Rainbow Bridge is a work in progress, and we are fortunate to stand here today when it seemingly floats in a state of consummation. But let's not tarry, let's move on to our second journey.

A GEOLOGIC TIME-LAPSE

Look downstream. Now let's speed up the passage of time and observe what happens. Days flutter by, years pass in a twinkling, centuries slip by in moments.

These stream- and wind-deposited sandstones are relatively soft. Viewed in time-lapse they melt away like ice cream. Bridge Creek occasionally floods. During big floods the normally-petite stream tumbles trees, bushes, unlucky rabbits, rocks, sand, and boulders. It whips around the canyon loops and batters the outside curves. In this way the loops grow wider, longer, and broader until they inevitably connect.

◁ **R**ainbow Bridge appears as a fin when viewed from the sweeping meander that Bridge Creek abandoned. The mound of rubble at the center of the curve is composed of stream cobbles stranded in the dry channel when Bridge Creek changed course.

Bridge Creek ▷ trickles through rock basins carved from bedrock. Unlike the wind-deposited Navajo Sandstone, the Kayenta Sandstone was emplaced in stream beds. It typically appears more purple than the overlying Navajo Sandstone and exhibits little of the spectacular cross-bedding patterns typical of the Navajo. It is slightly less permeable to water than the Navajo Sandstone. Thus, seeps often issue from the contact between the two formations.

⚠ **A**s ages passed and flash floods sporadically roared down from the mountaintop, the meanders of Bridge Creek looped larger and larger. Inevitably, the loops intersected. Bridge Creek followed the shortcuts. Now empty alcoves line both sides of the canyon and remnant loop partitions appear as barren knobs at canyon's center. Imagine Bridge Creek sweeping back and forth in an entrenched canyon until, one by one, each of the loop partitions broke through. As far as is known, only one of the breakthroughs preserved the cliff above the break. We call it Rainbow Bridge.

Soaring Sandstone

There are other natural bridges and arches in the world ▷ that approach or exceed the 275-foot span of Rainbow Bridge. Kachina Bridge in Natural Bridges National Monument spans 204 feet, Landscape Arch in Arches National Park reaches 291 feet, and Kolob Arch in Zion National Park holds the world's record for a freestanding arch. It vaults 310 feet! But no natural bridge or arch on Earth can rival the classic, sublime lines of Rainbow Bridge. The crest of Rainbow Bridge is 42 feet thick and 33 feet wide.

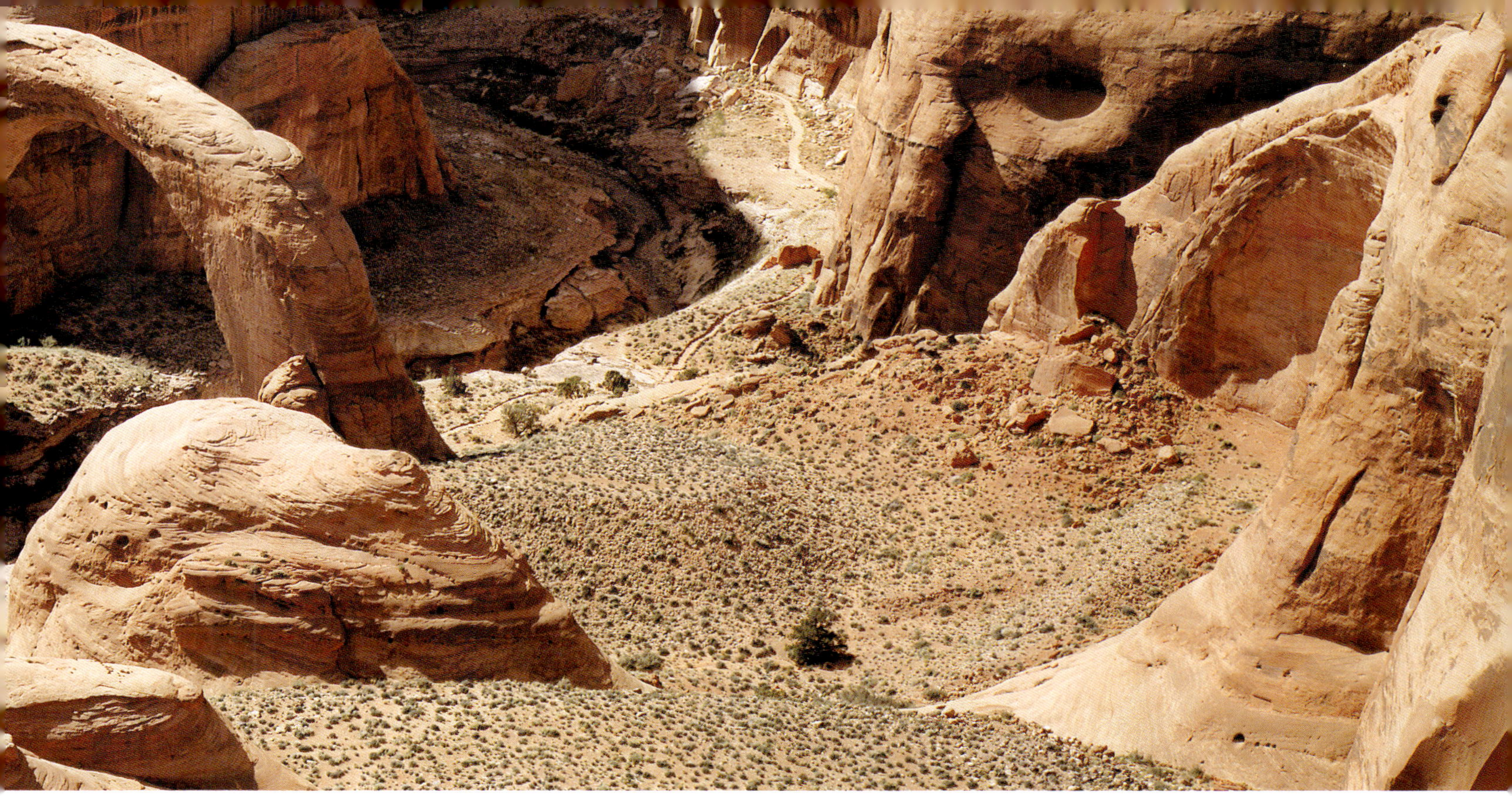

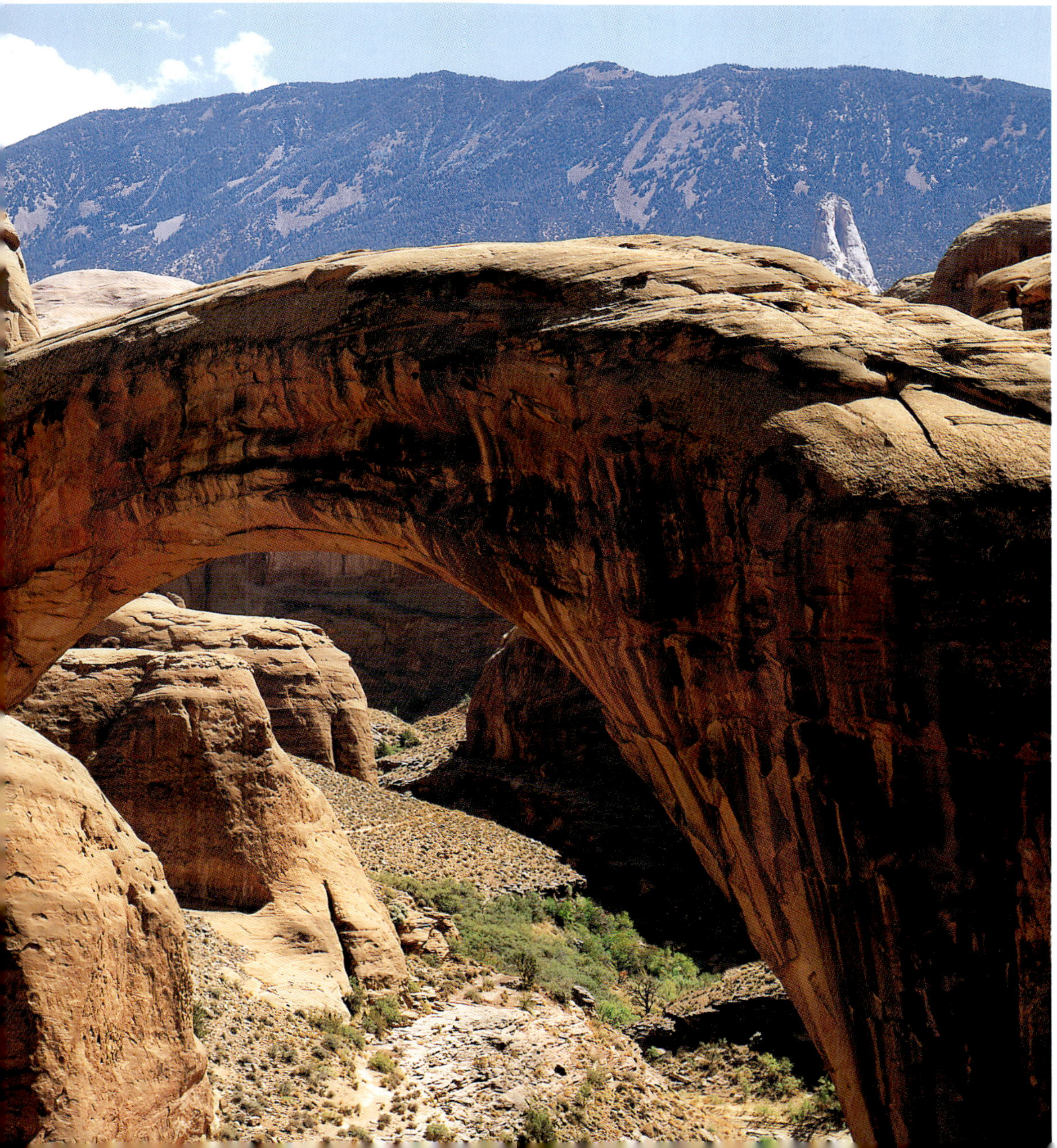

Forces of Water

Most of the time water quietly chisels away at the landscape, almost imperceptibly dissolving the cement that binds the rock particles or freezing and expanding in the cracks to break open the stone. But sometimes water turns into a savage brute. Storms bring sudden downpours. Waterfalls tumble from the cliff tops. Torrents gush from the mountain slopes. Tributaries combine their flows, and soon a deep rumble signals the approach of a flash flood. It is these brief periods when much of canyon-sculpturing takes place.

◁ **A splendid stream of sparkling water** cascades from the slickrock above Echo Camp in an alcove just upstream from Rainbow Bridge. Sudden downpours trigger brief waterfall spectacles. It was here at Echo Camp that a campsite was established for horseback riders approaching the bridge by trail.

An afternoon downpour over Navajo Mountain's Cliff Canyon brings rain, then waterfalls from the slickrock benches, then a roar and a flash flood. In a matter of minutes the depth of the flood can exceed six feet. Typically, such floods rage for an hour, then begin to subside. ▽

⚠ **D**awn comes clear and cold over Rainbow Bridge. A tattered blanket of snow covers the slickrock slopes of Navajo Mountain. Rainbow Bridge, with a height of 290 feet, is but a detail in a maze of sandstone canyons, domes, cliffs, benches, and hollows. Alcoves line the edge of Bridge Canyon, each undercut cliff a trace of an abandoned meander. It seems likely that the more-resistant Kayenta Sandstone slowed the downcutting of Bridge Creek. The stream, over centuries, skidded around on the surface of the harder rock. Most likely several meanders were abandoned near the contact between the Navajo Sandstone and the underlying Kayenta. All of the loops were lost, and now Bridge Creek flows fairly straight through an inner canyon rasped from the Kayenta while Rainbow Bridge is entirely sculptured from Navajo Sandstone.

A Span of Time

When Bridge Creek broke through a fin of rock, the breach was at first very small. As later floods widened the tunnel, other forces began to whittle away at the rock. We can see those powers at work here on the north leg of the bridge. Angular boulders lie at the base of the bridge where they have fallen from the underside of the span. More slabs will fall as cracks widen at both the urging of gravity and the insistence of water as it freezes in the cracks on long winter nights. The vertical and upper surfaces of the bridge are mostly rounded and smooth, reflecting the gentle action of weathering. These surfaces tend to slowly decompose grain by grain, or exfoliate, that is, break loose in thin sheets.

Discover America's Beauty
THE STORY BEHIND THE SCENERY

Enjoy—

Our National Parks

*The Culture of
our Native Americans*

*for those who then
want to know more*
We have a complete series of books

THE STORY BEHIND THE SCENERY®

Support our Park System

The Visitor Center has displays, information, and visual programs to help you better understand each unique area.

Associations and Concessionaires have other items in addition to our books to make your stay more enjoyable and rewarding.

Many areas have Field Institutes, Junior Ranger programs, and memberships for individuals and families.

Make your visit become— a Voyage of Discovery

Come explor. with u.

See how we present each subject in a colorful and entertaining way

See our *free* full-color catalog-

Call: 1-800-626-9673
        1-702-433-3415
Fax: 1-702-433-3420
Visit our web site:
<www.kcpublications.c

◁ **Was there once a bridge in this** nearby tributary of Forbidding Canyon? The evidence is tantalizing. A horn of rock rises from one side of the stream bed to point across to the other. Behind the horn, a sweeping alcove indicates that the stream once followed a circuitous loop. Someday the north leg of Rainbow Bridge may resemble this stump. Was this once a bridge leg? We'll never know for sure—we arrived a few thousand years too late.

The southern leg of ▷ Rainbow Bridge stands with a partner, a tower of rock on the upstream side. The monolith might be the remains of a broader Rainbow Bridge, unrefined by time and water. As future ages pass, Rainbow Bridge will most likely grow more slender and increasingly frail. Geologists use the term "joint controlled" to describe topographic features that are greatly influenced by ancient cracks. The crack between the bridge and the monolith helped to determine the width of the span.

*That evening
in the presence of growing doubt,
Nasja Begay rode into the firelight.
He said they were very close,
maybe a half day's ride away.*

In an Era of Discovery

It's 1909. Utah has been a state for 13 years. Nearby Arizona will remain a territory until 1912. Here, and also in New Mexico and Colorado, some astonishing geography has come to light in the desert canyons. Significantly, only 3 years ago Congress passed the Antiquities Act authorizing the president of the United States to set aside suitable public lands as national monuments.

THE SPLENDID DISCOVERIES

Richard Wetherill and Charlie Mason, cowboys from Mancos, Colorado, stumbled upon an imposing cliff dwelling in a cavernous alcove in southwest Colorado in December, 1888. They named it Cliff Palace. It was the first sighting of the scores of ruins that eventually led to the creation of Mesa Verde National Park in 1906.

In 1883 Cass Hite found a trio of massive sandstone bridges in Utah's White Canyon tributaries in what would become Natural Bridges National Monument in 1908. Richard Wetherill and Charlie Mason also discovered Keet Seel, the largest cliff dwelling in Arizona, in 1895; and 14 years later Richard's brother, John Wetherill, and archaeologist Byron Cummings came upon Betatakin, another majestic Anasazi Indian ruin near Keet Seel. Along with Inscription House, these spectacular ruins were incorporated into Navajo National Monument in 1909. Among numerous other discoveries were Aztec Ruins, Bandelier, and the Chaco Canyon ruins, all destined to be national monuments.

What other treasures might be sequestered in the canyons? What other astonishing archaeological sites awaited discovery on the mesas? The

▲ *In 1888, Richard Wetherill and his brother-in-law Charlie Mason stumbled upon a "magnificent city" in what would eventually become Mesa Verde National Park. This chance discovery was the spark that ignited a blazing passion for exploration in all five of the Wetherill brothers.*

Only one day after Cliff Palace and △
*then Spruce Tree House were discovered, Richard
Wetherill and Charlie Mason found another
imposing ruin that they soon called Square Tower
House. The two cowboys, at least for the
moment, forgot about their stray cows and
concentrated on the stone buildings and
their dusty artifacts. As would also be true of
Rainbow Bridge 21 years later, local Indians knew of
the ruins long before the Anglos' "discovery."*

Southwest seemed a place where the incredible
was probable. Explorers, archaeologists, and surveyors scrambled across the desert landscape,
looking for other revelations. Time was short for
these explorers because most of the country had
been thoroughly investigated. Yellowstone
National Park, the first national park, was
already 37 years old. The last frontiers were
rapidly vanishing.

THE WETHERILLS

When Richard Wetherill and his brother-
in-law, Charlie Mason, first glimpsed Cliff Palace
through the swirling snow, their lives and
the lives of their families as ranchers and cowboys were forever changed. The five sons of
Benjamin Wetherill—Richard, Al, John, Clayton,
and Win—all went on to pursue the treasures of
the Southwest.

John and his wife, Louisa Wade Wetherill, established a trading post on the Navajo Indian
Reservation at Oljeto, Utah, in 1906. From this
remote location, John Wetherill explored in all
directions, always pushing deeper into the

▲ **Sipapu Bridge, with a span of 268 feet**, *comes within 7 feet of equaling Rainbow Bridge. It is one of three huge stone bridges preserved in Natural Bridges National Monument. Both Cummings and Douglass surveyed this wondrous triad. Was it valid professional concern or annoyance with their overlapping interests that prompted Douglass to attempt revocation of Cummings's 1909 archaeological permits?*

wilderness. In addition to the local Indians, scientists, surveyors, daring tourists, and writers stopped by the trading post. It was a lonely outpost of civilization in a bewilderment of rock and sand. Wetherill's reputation as guide and outfitter had spread over the years, and now he was often hired to lead trips into the desert vastness.

Most accounts agree that Louisa Wetherill was the first to hear of a great natural bridge close to Navajo Mountain, from a Navajo Indian named Sharkie or One-Eyed Salt Clansman. Unfortunately, Sharkie died shortly thereafter.

Louisa and her explorer husband, however, were stirred by Sharkie's description. Where was this bridge if it, indeed, existed?

Byron Cummings

Byron Cummings, Professor of Ancient Languages and Literature and Dean at the University of Utah, conducted a survey of White Canyon's three bridges in the summer of 1907, a project that supported the creation of Natural Bridges National Monument the following year. In 1908 Professor Cummings was back again, this time to inspect the great cliff dwellings of the Tsegi and other nearby canyons. The Wetherill Trading Post at Oljeto became his headquarters. Most likely it was there that Cummings first heard the bridge rumors. Soon he and John Wetherill made plans to search for the bridge the following summer—if they could find someone who knew its location.

During the winter of 1908 or spring of 1909 two Paiute Indians responded to Louisa's inquiries about the elusive bridge. Nasja Begay and his father, Nasja, said they had been there and would be willing to serve as guides.

WILLIAM BOONE DOUGLASS

Meanwhile, in 1908, William B. Douglass, U.S. Examiner of Surveys, was surveying the natural bridges Cummings investigated the year before. His crew included a Paiute Indian known as Mike's Boy (later called Jim Mike). Mike's Boy seems also to have had some knowledge about the bridge, perhaps gained from a conversation with Nasja Begay, and he may have mentioned it to Douglass. Douglass reported the story to the General Land Office (the predecessor of today's Bureau of Land Management) for whom he was working and was subsequently instructed to investigate both the bridge and the ruins of Tsegi Canyon as possible national monument candidates.

Navajo National Monument, in northeastern Arizona, protects three imposing Anasazi Indian ruins. This one, Keet Seel, was first spotted by Richard and Al Wetherill, along with Charlie Mason, in 1895. It was John Wetherill and Byron Cummings who discovered both Betatakin (under the guidance of Navajo Clatzozen Benully) and Inscription House, only weeks before they went on to locate Rainbow Bridge, all in the unimaginably fruitful summer of 1909.

⚠️ ***The combined exploring parties of William Boone Douglass and Byron Cummings left John** and Louisa Wetherill's trading post at Oljeto, Utah, and passed through open country. The first night they camped near the towering hulk of Organ Rock. At this moment, no one could have guessed what the future would hold. Would John Wetherill have been amused to know that he would be appointed the first custodian of Rainbow Bridge National Monument?*

THE TRAIL AND TRIALS

In the summer of 1909 both Cummings and Douglass had research teams in the field and both had an interest in the big bridge. Douglass had attempted to find the bridge the previous winter but was foiled by deep snow and the failure of Mike's Boy to rendezvous as agreed. Now, Wetherill arranged for the groups to meet at Oljeto. From there they would travel westward toward Navajo Mountain, meeting Nasja Begay along the way.

The Cummings group of 7 included the dean's 11-year-old son, Malcolm. The Douglass group of 6 horsemen included Mike's Boy as their guide. With Nasja Begay the combined force would soon total 14. They started on August 10 and made their first camp a short distance from Oljeto near Organ Rock. There seems to have been some verbal sparring that night concerning which party had joined the other. Douglass especially, perhaps still smarting from his aborted trip the previous year, seemed anxious about his place in history.

The next morning Wetherill rousted the men from their sleep at 4:00 A.M. After breakfast they dropped into Copper Canyon, rounded the north end of No Mans Mesa, and turned southwest to camp in Nokai Canyon. The following day they ascended Piute Mesa, then rode on to the steep trail down a precipice into Piute Canyon.

When they reached Nasja's home in Piute Canyon they learned that Nasja Begay was away. Nasja said that he would send word for his son who would surely overtake the expedition soon.

No Mans Mesa forced the discovery party to ride far to the north close to the San Juan River. Kayenta Sandstone caps the mesa and its orange cliff of Wingate Sandstone. Like a colorful skirt, slopes of Chinle formation flare wide from the base of the Wingate cliffs. The Chinle formation is the same rock unit that bristles with petrified logs in ▽ Petrified Forest National Park.

Overleaf: Lake Powell slices ▷ across a stunning slickrock wilderness in the view northwest from the 10,388-foot summit of Navajo Mountain. If you have sharp eyes, you'll be able to pick out tiny Rainbow Bridge soaring across its canyon near the panorama's left edge.

△ **A backpacker follows the trail** to Rainbow Bridge in Bald Rock Canyon. The discovery party termed the barren Navajo Sandstone domes the "baldheads."

At the end of the fourth day, ▷ beyond the reach of regular trails, the discovery party entered a rock-bound vale. They camped here and named it Surprise Valley. But Nasja Begay, the Indian guide who knew the way to the phantom bridge, remained unaccounted for. After a dinner of rice, canned corn, biscuits, and tea, Nasja Begay rode in from the dark. He announced that the bridge was a half-day's ride away.

The horsemen continued into the canyon maze. That night their third camp was made in Cha Canyon near today's trailhead.

Wetherill may have been in this vicinity in 1907 but from here on the country seemed to be unknown to him. They were on the threshold of a tangled complex of canyons and mesas that had long shielded the rumored bridge from the outside world.

INTO THE MAZE

The following morning the expedition encountered difficulty finding a route through the sandstone domes. Mike's Boy seemed confused and apparently wanted to turn back. Cummings later recounted that even Douglass voiced a desire to quit.

The slickrock terrain showed no trace of a trail and made the footing treacherous for the animals. The men dismounted on the steepest grades. Finally they came to what they called Surprise Valley. Here they made camp number four. Every horse was suffering from the punishment of the route; several had bleeding hooves. Yet the straight-line distance covered for the day's work measured little more than three miles. Where was Nasja Begay?

△ **Sunset splashes gold on the "glass mountains"** at the entrance to Surprise Valley. Zane Grey described the sandstone billows as glass mountains and featured Surprise Valley as a fateful location in the novel The Rainbow Trail.

Cha Canyon's Beaver ▷
Creek reflects a morning sky near the discovery party's third camp. It was in this area that the Indian trails disappeared. Wetherill's business partner, Clyde Colville, and a Navajo guide explored this far in 1908 in search of the fabled bridge, but the maze of sandstone domes beyond thwarted further penetration.

▲ **O**wl Bridge is technically an arch since it doesn't span a watercourse. The Cummings-Douglass party spotted and photographed it at the edge of Surprise Valley the morning of the day they discovered Rainbow Bridge.

Oak Canyon twists down the northwest flank of Navajo Mountain at the near ▷ left while Bridge Canyon cleaves the slickrock slopes on the far right. It was through this tangled maze of canyons and cracks that the discovery party broke trail to Rainbow Bridge on the final day. The domed profile of Navajo Mountain hints at its laccolithic birth when magma welled up from below to warp the overlying sedimentary layers into a blister-shaped peak.

Neil M. Judd, a nephew of Cummings and destined to become a respected archaeologist in his own right, said, "our Indians [Mike's Boy and Dogeye Begay, a Navajo wrangler and cook]…were in a dither. Both wanted to turn back; they had gone far enough. We were lost! There was no escape from these infernal gorges."

To their left, the rugged palisades and forested summit of Navajo Mountain loomed a mile above them. To their right, the dark San Juan River gorge twisted across the slickrock desert. Ahead of them, a series of canyons tumbled from the mountain toward the hidden river. Waves of civilization had flowed around this wilderness stronghold. Now an expedition in search of a stone rainbow was poised to force an opening.

That evening in the presence of growing doubt, Nasja Begay rode into the firelight. He said they were very close, maybe a half day's ride away.

THE FINAL MILES

The next morning they passed Owl Bridge (actually an arch) and may have spotted White Crag Arch high on the slopes of Navajo Mountain. They surmounted the divide between Surprise Valley and Oak Canyon's Paradise Valley. Another narrow pass led to a descent into what they called "Hidden Valley." After an early lunch they proceeded down Nonnezoshie Boko (Bridge Canyon), the final leg of their quest.

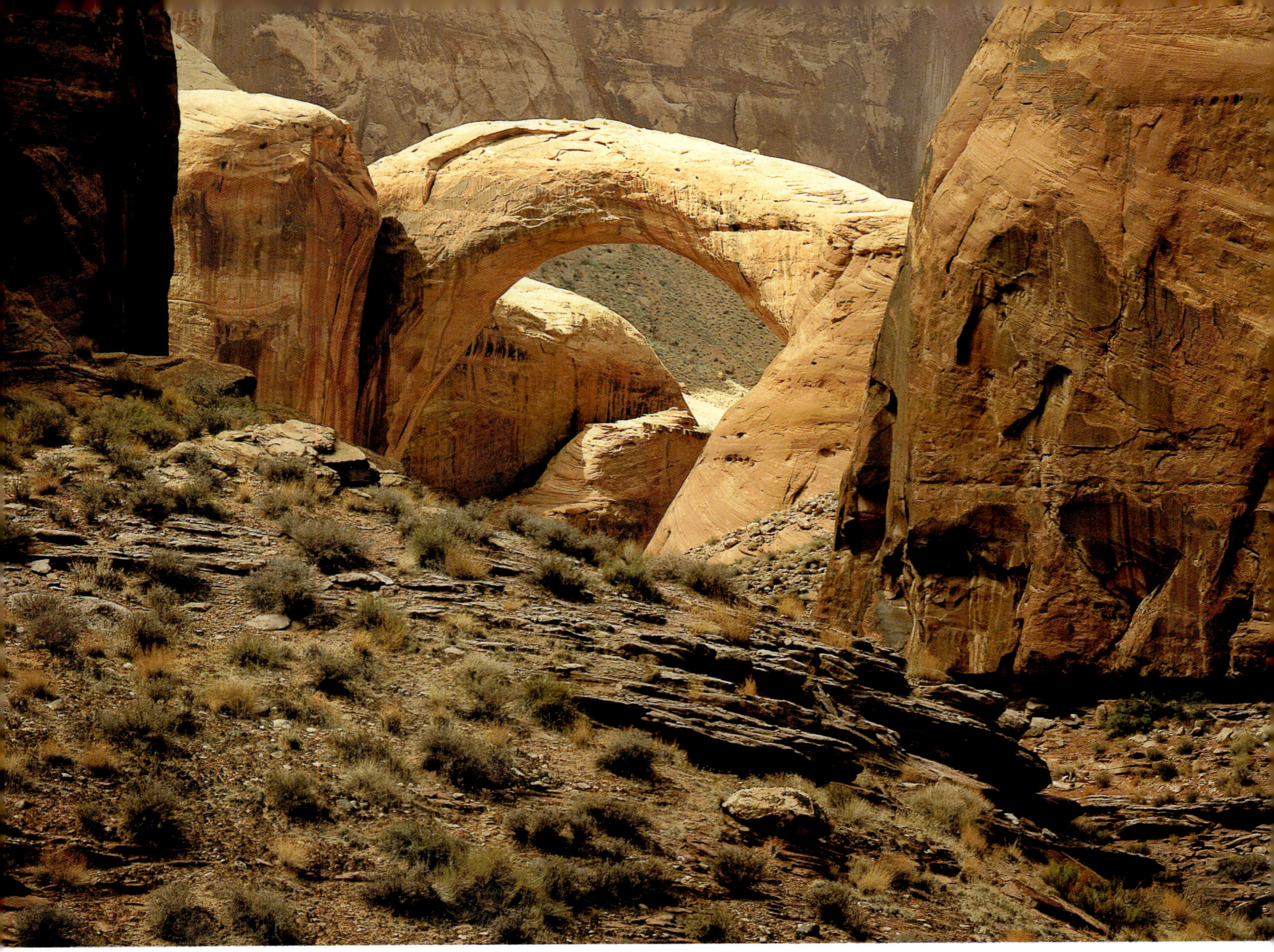

It was from this point in the depths of *Nonnezoshie Boko (Bridge Canyon) that the discovery party first glimpsed their goal. From the rim of the canyon, today's trail is apparent as it follows the Kayenta Sandstone bench toward its entry into the national monument. It is believed that John Wetherill was the first of the expedition's members to pass beneath the bridge, followed closely by William Douglass and Byron Cummings. But the journals and interview transcripts do not absolutely concur on such details, and it is mystifying that the expedition accounts so often disagree. Were these men really on the same trip?*

Several accounts of these final miles relate that Douglass seemed fixed on a first sighting. In his effort to stay ahead, Douglass's horsemanship became a danger to his mount and an aggravation to his colleagues. Because of a peculiarity in the course of the canyon just upstream from the bridge (the remnants of an ancient abandoned meander), the first viewpoint is gained by looking to the extreme left and behind. Cummings, heeding Nasja Begay's advice, spotted the bridge first. Wetherill became the first to pass beneath it. Malcolm Cummings claimed he was dead last. Not everyone, Douglass in particular, agreed who did what first. Nevertheless, whatever happened, it was midday, August 14, 1909.

John Wetherill climbed cliffs and *scrambled up a ravine to leave a record of the discovery at the base of a towering sandstone wall.*

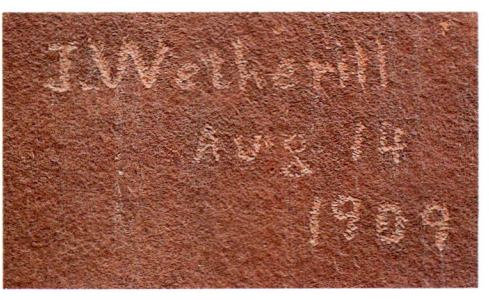

Junipers and ▷ an oak glen crowd a cliff in Bridge Canyon along the Rainbow Bridge Trail while fingers of desert varnish reach down from the cliff tops. Even without a Rainbow Bridge, the canyon habitat is a treasure all its own.

◁ *Ricegrass* dominates the plant community on a rugged slope close to Rainbow Bridge. Although the great bridge draws our attention like a magnet, it can also be said that its setting of grasses, junipers, flowers, cobbles, and patina-streaked alcoves, is equally important to the site's harmony and beauty.

Mr.
Mrs.
Ms. _____
         First Name   *(Please Print)*   *Last Name*

Address _____

City _____ State/Province _____

Zip _____ Country _____

Day Phone _____

Fax _____ E-Mail _____

❏ Personal Use   ❏ Retail   ❏ Educational

Comments _____

The remainder of that glorious day was filled with work, cooling off in the pools of Bridge Creek, and reconnoitering. Six of the Cummings party walked five miles to the Colorado River and back again while others explored and surveyed the immediate area. A name for the new bridge was a point of debate. *Nonnezoshie*, roughly translated from the Navajo, means "bridge." *Barahoini*, the Paiute word for "rainbow," was suggested. Also, the Paiute word for "Space Under a Horse's Belly" was considered…perhaps only briefly! "Rainbow Bridge" soon eased into the vocabulary of the discoverers.

Most of the Cummings party left for Oljeto the following day after giving Douglass a portion of their provisions. Douglass, his team, and two men from the Cummings group stayed all or part of an additional four days to locate the 160 acres that would, the following year, encompass the national monument. Then they moved on to the Tsegi canyons, leaving the great stone rainbow in peace. Both groups went hungry before returning to civilization.

Forbidding Canyon normally plays host to a gentle Aztec Creek quietly slipping from pool to pool. But even at low flows, evidence of deluges abounds. Rippled sandbars indicate a recent flood, while logs and tree limbs caught on a ledge tell a tale of rumbling waters several feet deep.

White Crag Arch stands 1,400 feet above the trail to Rainbow Bridge. Like Rainbow Bridge, White Crag is carved from Navajo Sandstone.

*Prior to the creation of Lake Powell
only a few tens of thousands of people
visited Rainbow Bridge.
Today, about 300,000 people come to look
upon the rainbow-turned-to-stone each year.*

Continuing Discovery

It is ironic that the "discoverers" of Rainbow Bridge were, in fact, newcomers. Their Indian guides and surely other Native Americans knew the bridge. In nearby canyons there are Anglo inscriptions that predate the Cummings-Douglass expedition by decades. Judd mentioned that in their dash down to the Colorado River on the afternoon of the discovery they found camp equipment, miners' implements, and other evi-

▲ **It's probable that Rainbow Bridge was** discovered by Anglos long before Cummings and Douglass found it. Miners left names and dates in Forbidding Canyon in the early 1880s.

▲ **Before Lake Powell allowed easy** access to Rainbow Bridge, visitors approached by horseback. Echo Camp was established with tents, beds, and a chuck box to help riders feel at home.

dence of Anglo visitation. (A gold rush of sorts occurred in Glen Canyon around the turn of the century, and it is nearly impossible to imagine that not a single prospector probed up Bridge Canyon.)

A heliograph station, operated by the U.S. Army, was active on the summit of Navajo Mountain in 1864 and the bridge is visible 6,700 feet below and 5½ miles distant. Did not any of the soldiers discern the shape in the confusion of sandstone canyons?

Furthermore, the Anasazi Indians lived in the region for hundreds of years until about A.D. 1300. Carbon dating indicates that constructed fire pits near the bridge are close to 2,000 years old.

△ ▷ **P**rior to Columbus's discovery of America, Anasazi Indians lived along the Colorado River and its
tributaries. In the vicinity of Bridge Canyon, pictographs (rock paintings), petroglyphs (pecked rock figures),
food storage bins, toehold routes, pottery shards, fire pits, and dwellings dot the canyons and mesas. The
Anasazi and their ancestors dwelt here for over a thousand years before they suddenly moved away about A.D.
1300. The true discoverers of Rainbow Bridge will never be known.

As must happen often, the so-called discoverers are those who first catch the attention of the world. The actual discoverers are lost in the mists of history. Douglass need not have been so feverish! But it's not difficult to sympathize with the government man. Most of us wish to link with the stone rainbow by being the first in our group to spot it, by photographing it, or by following the stars as they arc behind its graceful silhouette.

CHARLES L. BERNHEIMER

Newspaper articles proclaiming the bridge soon appeared. *National Geographic Magazine* published an article in 1911. Theodore Roosevelt and Zane Grey were among the first 60 Anglos to view the colossal bridge. But even after the veil of the unknown had been withdrawn, sheer isolation kept visitation sporadic. Ten years after discovery, the visitor register at the bridge, which may not be entirely complete, recorded fewer than 200 signatures.

A brief gold rush occurred in Glen Canyon in ▷
the 1890s. These steps were chopped into a sandstone
slope to help pack animals negotiate the route.

One man, however, irresistibly drawn to the bridge was a self-described "tenderfoot and cliff dweller in Manhattan." Charles L. Bernheimer, a wealthy businessman and labor negotiator, was born in Germany and educated in Geneva. Bernheimer was frail in appearance and an unlikely figure for the hardships of desert trails. Yet he made 15 rugged trips into the desert Southwest. Guided by John Wetherill, four of these journeys (in 1920, 1921, 1922, and 1924) focused on the Rainbow Plateau.

All of the Bernheimer expeditions penetrated canyons previously unknown to white men. Many landscape features—Forbidding Canyon (earlier called West Canyon), No Name Mesa, Cliff Canyon, and Redbud Pass—were named during these exploits. But the 1922 trip was especially

▲ **For decades John Wetherill continued to** explore the region. Charlie Mason, co-discoverer of Mesa Verde National Park's Cliff Palace along with John's brother, Richard, was a frequent companion.

This ragged slash in ▷ the side of Navajo Mountain is the head of Cliff Canyon. In 1922 when John Wetherill, famed archaeologist Earl Morris, first Natural Bridges National Monument Custodian Zeke Johnson, New Yorker Charles Bernheimer, and others forged a second trail to Rainbow Bridge, they descended this defile with pack animals 1,800 feet from a notch between Navajo Mountain and what they called No Name Mesa. Even today, backpackers test their knees on the difficult descent to the bed of Cliff Canyon. It leads to a crack that is the key element along the Bernheimer Trail—Redbud Pass.

◁ *An affection for the landscape and a passion for discovery fired John Wetherill's explorations just as they do for many of today's backpackers. In 1921 Bernheimer wrote that "Wetherill's memory of the faint and intermittent Indian trails is most remarkable. We strayed but rarely. He is beyond all doubt the typical pathfinder or pathmaker."*

significant. The primary goal was to locate an approach route to Rainbow Bridge from the south, from the far side of Navajo Mountain. Earlier attempts, working downstream in Forbidding Canyon, were stymied by cliffs and pouroffs.

The 1922 explorers hit upon a route that bypassed Forbidding Canyon. A high pass between No Name Mesa and the mountain led to a precipitous 1,800-foot descent into Cliff Canyon. Searching downstream they were again thwarted by pouroffs. Then, checking into tributary cracks feeding Cliff Canyon from the north, Wetherill found a connection to a Bridge Canyon tributary.

The route, however, needed a little "persuasion" to make it passable to pack animals. Bernheimer and Wetherill came prepared: along with sledges, shovels, picks, and drills, they brought a load of "dynamite, TNT and black powder"! The persuasion was successful and they soon squeezed through Redbud Pass on the new "Bernheimer Trail" to Rainbow Bridge.

Redbuds prosper while backpackers ▷ struggle through a crack along an obscure route to Rainbow Bridge. These ramblers have forsaken the trail to follow dubious "shortcuts" through the slickrock maze of canyons and cracks that surround Navajo Mountain. The sport of getting lost in the canyons thrives in the tradition of Cummings, Wetherill, Douglass, Bernheimer— and Teddy Roosevelt, who visited the bridge in 1913.

If Cummings, Douglass, Wetherill, Nasja Begay, and the others rode down Nonnezoshie Boko to the first viewpoint today, Cummings might well insist that Mr. Douglass take the lead. From the overlook their observations would most likely find Rainbow Bridge unchanged.

Downstream, however, they would find dramatic transformations. Lake Powell began to fill in 1963. Seventeen years later the full lake extended into the national monument, under and well beyond the bridge. Gone are the seeps, springs, maidenhair ferns, and monkey flowers that thrived along the stream course. Instead, there are often astonishing reflections of the soaring bridge and its canyon setting.

Prior to the creation of Lake Powell only a few tens of thousands of people visited Rainbow Bridge. Today, because of the easy access afforded by the lake, about 300,000 people come to look upon the rainbow-turned-to-stone each year. Cummings, Douglass, Bernheimer, and Wetherill would be flabbergasted.

Lake Powell with its power boats, docks, and floating walkways is a striking departure from the 1909 bridge of discovery. When the lake level falls, thick silt beds are revealed. When the lake rises, the engaging reflections notwithstanding, an artificial inundation occurs.

To protect the national monument from the rising lake, a pair of dams within Bridge Canyon was proposed. One dam would have been built downstream to hold back Lake Powell. Upstream, another dam would have diverted Bridge Creek into a tunnel that would carry it to Forbidding Canyon. Fortunately, these projects were not realized. Their access roads, power lines, construction activity, dams, and tunnel would have inflicted far more harm than the encroachment of the lake.

The National Park Service is mandated to retain the bridge much as it was in August, 1909. It is not an easy assignment. The presence of the lake and the crush of admirers strains the natural environment. Trails have been built to assist visitors and to prevent trampling of the sparse desert vegetation. Visitors are asked to maintain a serene atmosphere befitting a revered natural wonder— but sometimes, especially when the weather is wild and no one is around, it's difficult to avoid letting go with a hair-raising "Whoop!"

▲ *Until 1924, when the Rainbow Lodge opened, John Wetherill led nearly all of the expeditions to Rainbow Bridge. The typical trip began at the Wetherill Trading Post (which was moved from Oljeto, Utah, to Kayenta, Arizona, in 1910), circled through the Tsegi canyons, connected with the discovery party's 1909 route, then turned back after a brief visit to the great bridge. The trip required eight days. Most of today's visitors arrive by boat via Lake Powell, less than two hours run from Wahweap and Bullfrog marinas.*

_V_isitors admire the bridge △
from a rock-floored viewing
area. During his epic exploration
trips of 1869 and 1871, J. W.
Powell boated by the mouth of
Forbidding Canyon never
knowing what lay five miles away.

◁ **_A_ rising Lake Powell surrounds**
a sacred datura plant in early July.
The large white flowers of the datura
open at night, then wilt in morning's
sunlight. Beware! All parts of the
plant are poisonous.

Floral Colors in the Canyons

The barren slickrock desert can be hostile during the summer months when temperatures often reach a hundred degrees or more. But in the spring when temperatures are moderate and snowmelt feeds the creeks, the flowers can turn the canyons into lovely gardens. April and May offer the greatest rewards for those who hike the canyons and explore the plateaus. The show lasts a few weeks before seasonal heat brings down the curtain.

◁ **Prickly pear and rough mules ears** greet hikers at the southern entrance to Redbud Pass. It was here that the Bernheimer Expedition camped in 1922.

△ **Sand verbena blooms open** in the early evening to fill the night air with sweet fragrance.

◁ **The green stems of Mormon tea** can be used to brew a pleasant hot drink. It contains traces of the drug ephedrine, useful as a muscle relaxant.

△ **An early summer bouquet of spiderwort**
brightens the Rainbow Bridge Trail near the
confluence of Redbud Canyon and Bridge Canyon.
Spiderworts open in the morning and close again
by early afternoon. They grow most often in sandy
soils shaded by cliffs or other plants.

△ **The brilliant color of Indian paintbrush**
lights the way along the discovery trail
at Cha Canyon.

◁ **Cottonwood trees**
usually indicate a
dependable water
supply. The shade they
provide is a welcome
haven for canyon travelers.
The broad leaves of
the cottonwood resemble
those of the aspen tree,
to which it is related,
but cottonwoods prefer
much lower elevations.
This one thrives on
a sandy bench in
Nasja Canyon a short
distance downstream
from Surprise Valley.

41

In 1924, only ▷
two years after the Bernheimer Trail was established, a road was pushed to Navajo Mountain and a lodge built at the effective head of the trail. Today, the lodge and trading post lie in ruins while Cummings Mesa looms on the distant horizon.

◁ *While on a Norman Nevills Colorado River float trip in 1940, Barry Goldwater hiked up Forbidding and Bridge canyons to visit a place of "intense interest" to him, Rainbow Bridge. Enthralled by the wild beauty of the sandstone canyons, Goldwater soon purchased the Rainbow Lodge at the head of the overland route to the magnificent bridge.*

⚠ **Lightning bolts stab the Navajo Indian Reservation south of the Rainbow Lodge ruins. The main** lodge building burned in 1951. Unoccupied from 1951-1958, the remaining buildings were used for the last time in 1965. By some counts, a total of about 25,000 people had visited the remote bridge by 1963 when Lake Powell began to fill.

A Land of Mystery and Majesty

⚠ **_Surprises are commonplace in the_**
sandstone landscapes of southern Utah.
If a photograph of Rainbow Bridge is printed
upside down, who will notice in this land of
wonders? A radical idea that could turn the
entire discovery story upside down is a hint
that Wetherill visited Rainbow Bridge in
1908! Revealingly, Wetherill gave credit to the
Bernheimer Expedition for finding Redbud
Pass even though he had carved his name
there 11 years earlier! Rumors persist that he
also had foreknowledge of Rainbow Bridge.

SUGGESTED READING

BABBITT, JAMES E., ed. *Rainbow Trails: Early-day Adventures in Rainbow Bridge Country.* Page, Arizona: Glen Canyon Natural History Association, 1990.

BERNHEIMER, CHARLES L. *Rainbow Bridge.* New York: Doubleday, 1926.

JETT, STEPHEN C. "The Great 'Race' to 'Discover' Rainbow Natural Bridge in 1909." *Kiva.* Vol. 58, No. 1. Tucson, Arizona: Arizona Archaeological and Historical Society, Inc., 1992.

LEAKE, HARVEY and GARY TOPPING. "The Bernheimer Explorations in Forbidding Canyon." *Utah Historical Quarterly.* Volume 55, Number 2. Salt Lake City, Utah: Utah State Historical Society, 1987.

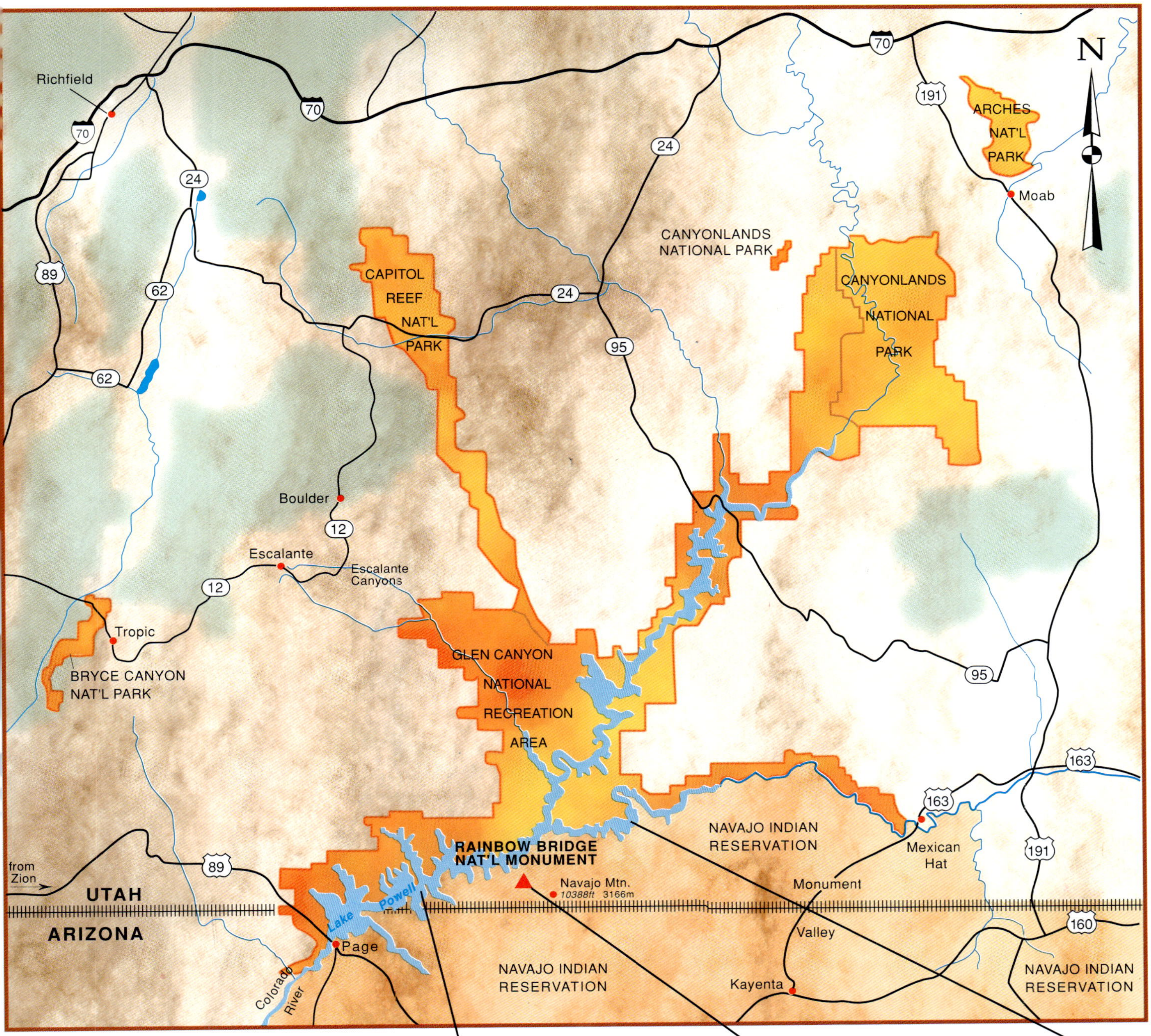

N

Richfield

70

24

89

62

62

24

CAPITOL REEF NAT'L PARK

CANYONLANDS NATIONAL PARK

CANYONLANDS NATIONAL PARK

191

ARCHES NAT'L PARK

Moab

24

95

Boulder

12

Escalante

Escalante Canyons

12

Tropic

BRYCE CANYON NAT'L PARK

GLEN CANYON NATIONAL RECREATION AREA

95

163

163

NAVAJO INDIAN RESERVATION

Mexican Hat

191

89

from Zion →

UTAH

ARIZONA

Lake Powell

RAINBOW BRIDGE NAT'L MONUMENT

Navajo Mtn.
10388ft 3166m

Monument

Valley

160

Page

Colorado River

NAVAJO INDIAN RESERVATION

Kayenta

NAVAJO INDIAN RESERVATION

ACCESS: Rainbow Bridge National Monument is most often reached by boat on Lake Powell. Tour boat service is available from Wahweap, Bullfrog, and Halls Crossing marinas; rental or private boats leave from any marina except Dangling Rope. Hikers may reach the monument via rough roads and either of two lengthy, rugged, non-maintained trails. Each trail requires a Navajo Nation hiking and camping permit. Scenic air tours operate out of Page, Arizona.

WEATHER: Summer temperatures can soar to over 100˚F making trail access dangerous. Distant downpours can cause flash floods in narrow canyons. Winter often brings sub-freezing temperatures to the monument.

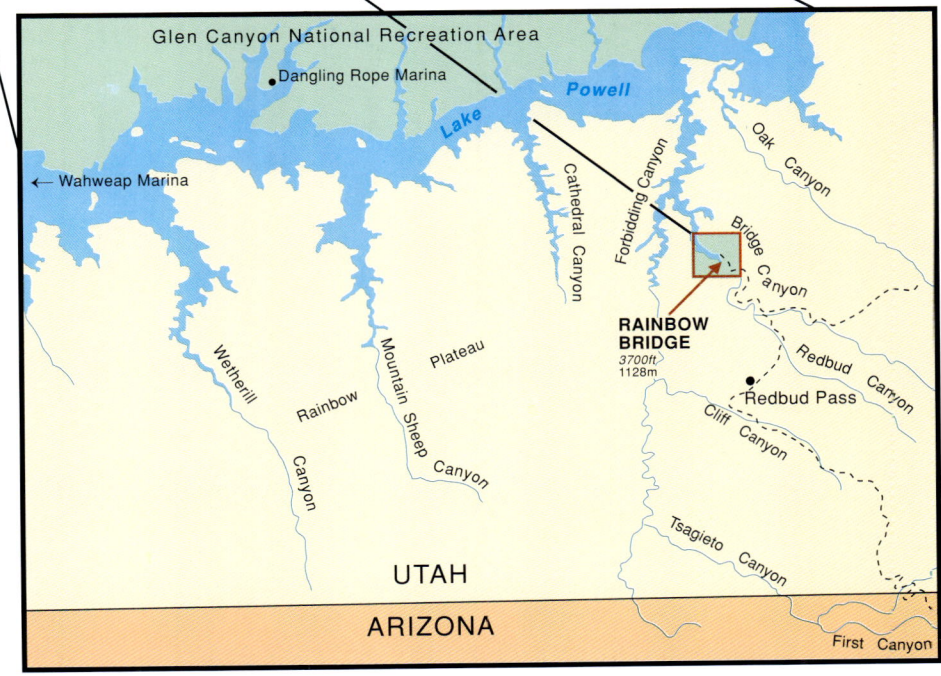

Glen Canyon National Recreation Area

Dangling Rope Marina

Lake Powell

← Wahweap Marina

Wetherill Canyon

Rainbow Plateau

Mountain Sheep Canyon

Cathedral Canyon

Forbidding Canyon

Oak Canyon

Bridge Canyon

RAINBOW BRIDGE
3700ft
1128m

Redbud Canyon

Redbud Pass

Cliff Canyon

Tsagieto Canyon

UTAH

ARIZONA

First Canyon

*F*ew people visit Rainbow Bridge in mid-winter when sunlight skims across the mesas to ricochet back and forth across the tranquil canyon. It is a scene reminiscent of centuries of isolation. In January it is not difficult to imagine Anasazi Indians walking with quiet respect toward the span, a miner with pick and shovel in hand gazing transfixed, Byron Cummings making notes, or William Boone Douglass calculating the bridge's dimensions. Less than a year after its discovery, on May 30, 1910, President William H. Taft created Rainbow Bridge National Monument. Visitors have been awed by this wonder of the world ever since. Zane Grey described the bridge as "…black and mystic at night, transparent and rosy in the sunrise, at sunset a flaming curve limned against the heavens."

The trail from the courtesy docks to the bridge viewpoint meanders along the edge of Lake Powell. Boulders, fallen from a nearby alcove, remind us of the forces that carved the rainbow-shaped bridge.

Rainbow Bridge: The Miracle

Rainbow Bridge is a remarkable landform. It evokes many emotions—wonder, awe, respect, pleasure. Several Southwestern Native American tribes hold it in reverence. Rainbow Bridge is not a mere geological curiosity. It is a place to salute all the slickrock terrain, including that lost beneath Lake Powell. Rainbow Bridge, like our Planet Earth, lives, changes, and matures.

One winter, when few visitors were about, I walked to the span. There on the ground was a fallen fragment of Rainbow Bridge. The long night's freezing temperatures had wedged off a slab of Navajo Sandstone. It fell a hundred feet and smashed on the Kayenta Sandstone bench. I picked up the largest fragment. It was weathered and partially mantled in desert varnish. Philosophically I was not surprised by the collapse. Emotionally I was stunned. Yes, the ages continue to refine Rainbow Bridge.

The bridge conveys a convincing illusion of permanence. But in geologic time, it is ephemeral. How long will Rainbow Bridge stand? We have not a clue.

Rainbow Bridge is a jewel in a setting of gold and silver: It captures the imagination as no other single topographic expression. Yet it is but a minor detail in a landscape studded with strange and astonishing features. Consider those elements that are NOT Rainbow Bridge—the curve of the cliffs, the desert varnish, the shaggy junipers, sacred datura blossoms, and little Bridge Creek. All of them are miracles in a tiny square of land lost in a slickrock labyrinth.

Books on National Park areas in "The Story Behind the Scenery" series are: Acadia, Alcatraz Island, Arches, Badlands, Big Bend, Biscayne, Blue Ridge Parkway, Bryce Canyon, Canyon de Chelly, Canyonlands, Cape Cod, Capitol Reef, Channel Islands, Civil War Parks, Colonial, Crater Lake, Death Valley, Denali, Devils Tower, Dinosaur, Everglades, Fort Clatsop, Gettysburg, Glacier, Glen Canyon-Lake Powell, Grand Canyon, Grand Canyon-North Rim, Grand Teton, Great Basin, Great Smoky Mountains, Haleakalā, Hawai`i Volcanoes, Independence, Jewel Cave, Joshua Tree, Lake Mead-Hoover Dam, Lassen Volcanic, Lincoln Parks, Mammoth Cave, Mesa Verde, Mount Rainier, Mount Rushmore, Mount St. Helens, National Park Service, National Seashores, North Cascades, Olympic, Petrified Forest, Rainbow Bridge, Redwood, Rocky Mountain, Scotty's Castle, Sequoia & Kings Canyon, Shenandoah, Statue of Liberty, Theodore Roosevelt, Virgin Islands, Wind Cave, Yellowstone, Yosemite, Zion.

A companion series on National Park areas is the "in pictures...The Continuing Story." This series has **Translation Packages**, providing each title with a complete text both in English and, individually, a second language, German, French, or Japanese. Selected titles in both this series and our other books are available in up to 8 languages.

Additional books in "The Story Behind the Scenery" series are: Annapolis, Big Sur, California Gold Country, California Trail, Colorado Plateau, Columbia River Gorge, Fire: A Force of Nature, Grand Circle Adventure, John Wesley Powell, Kauai, Lake Tahoe, Las Vegas, Lewis & Clark, Monument Valley, Mormon Temple Square, Mormon Trail, Mount St. Helens, Nevada's Red Rock Canyon, Nevada's Valley of Fire, Oregon Trail, Oregon Trail Center, Santa Catalina, Santa Fe Trail, Sharks, Sonoran Desert, U.S. Virgin Islands, Water: A Gift of Nature, Whales.

Call (800-626-9673), fax (702-433-3420), write to the address below,
   Or visit our website at www.kcpublications.com

Published by KC Publications, 3245 E. Patrick Ln., Suite A, Las Vegas, NV 89120.

Inside back cover: "...as I floated on my back in the water, the Bridge towered above me."
—Teddy Roosevelt.

Created, Designed, and Published in the U.S.A.
Printed by Doosan Dong-A Co., Ltd., Seoul, Korea
Color Separations by Kedia/Kwang Yang Sa Co., Ltd.
Paper produced exclusively by Hankuk Paper Mfg. Co., Ltd.